John E. Powers

Reducing Friction

Elibron Classics
www.elibron.com

Elibron Classics series.

ISBN 0-543-70249-9 (paperback)
ISBN 0-543-70248-0 (hardcover)

This Elibron Classics Replica Edition is an unabridged facsimile
of the edition published in 1899 by Vacuum Oil Company, London.

REDUCING FRICTION

BY JOHN E POWERS

1899

VACUUM OIL COMPANY
LONDON

CONTENTS

One oil lubricates more than another

REDUCING FRICTION

I

THE GIST OF THE BOOK

This book is written to show that the value of oil to a user of power—the price per gallon that he can afford to pay—depends on the answers to these questions:

Where to look for the value of oil

(1) how easy does it make his machinery go?

(2) how much power does it save?

(3) how much is this savable power worth in his particular works?—It differs in different circumstances. It costs all the way from nothing to $100 a horse-power—more perhaps.

(4) how much wear does the oil prevent?

(5) how much do preventable repairs cost?

(6) how much, besides cost of repairs, is his loss by preventable wear?

(7) how long does the oil itself last? in the sense of: how much has he got to buy in a month or year?

There are no other questions to ask about oil itself, except, in some businesses, special questions like these: does it stain? does it fly off? etc.

Each chapter contains one part of the subject: one step in the journey. The end of the journey is such answers to those seven questions as can be put in a book. That is oil-intelligence.

Books and experience

Books alone do not make engineers mechanics or mill-men; books and experience make them better than either alone. The use of oil is best seen in a book; but the uses of oil must be learned in the shop.

II

OIL-INTELLIGENCE

There is money in knowledge of oil

The object is to show a user of power that oil is a money-saver and -maker, whether he uses it well or not; but that, if he uses it well, it saves and makes a good deal of money for him.*

* Professor Thurston says that the art of using oil consists of choosing the one that best suits each place, and applying it in the way that best suits its place.

Robert H Thurston, LL D, is Director of Sibley College, the Schools of Mechanical Engineering and of the Mechanic Arts, Cornell University, Ithaca, N Y, U S A, and author of *Friction and Lost Work in Machinery and Mill Work* (New York: John Wiley & Sons, $3. London: Trübner & Co, 12s); from which we shall quote quite freely. There is no higher authority.

We venture to advise the reader, willing to search through a technical book for parts that he can read, to buy

Intelligence pays

Uses it well—what do we mean? When you speak of a man and a tool, and say that he uses it well, you mean that he knows his work with that tool. We mean: if you know your work with oil, it is worth a good deal of money to you. Some parts of machinery can be worked with almost any oil; some parts can scarcely be worked at all without fairly suitable oil. All parts work best with oils that suit them. Oil-intelligence pays.

Book and shop

What is this oil-intelligence? We know oil, and you know your work; neither one of us knows the particular oils that are best for the several parts of your work—just that is the oil-intelligence you require; and nobody has it. It cannot be put in a book.

From the book

There is a way to get it, though, near enough. Learn what oil is for; what it does; how it does it. What! are we saying that you don't know these simple things? They are not so very simple; we try to make them plain.

the book above-mentioned. It is a text-book in mechanical engineering: not to be mastered without the higher mathematics. But we know of no more accessible confirmation of our own publications.

We consult the Author professionally, in our manufacturing, and translate his learned books into *a b c*, as well as we can, for popular reading.

What oil-intelligence is

Oil-intelligence takes in coal, repairs, detentions, breakdowns, the life of machinery. Oil affects them all: your choice of oil affects them all. There are great lack and need of oil-intelligence.

True

This is our advertisement; but our object is not to sell a particular oil. We probably make all the oils you need. We teach lubrication, as if we were teacher and maker only, not seller at all. We have no partialities.

III

INTELLIGENT OIL-MANUFACTURING

Oil makes ease

Oil makes machinery go with ease—we all know it.

Oils differ

Oils differ: one gives more ease than another—we all know that.

Machinery differs

Machinery differs: one part of a machine is more eased with this oil; another with that—we all agree, so far.

The oil that lubricates most

It follows: there is a possible oil for every friction-place that will give more ease, at that place, than any other oil. We agree to this too; but some of us think it spun too fine. And it is; but it points to a practical fact of the greatest importance. This:

A maker of oils should consider all parts of all machinery, group them as friction-producers in twenty or thirty groups perhaps, make an oil for each group: an oil to give more ease to its group than any other possible oil.

Friction-places grouped

An oil for every group

The only qualification to this is the varying friction in apparently-similar places, due to faults in the metal make or adjustment. Such variables, sometimes, control the oil-requirements. Correct the fault, and the place falls into its group. That is: the place itself was out of its group. There is no exception.

Exceptions

IV

KNOWING ABOUT LUBRICATION AND OIL

There is a whole set of knowledge about lubrication and oil. A common man has some of this knowledge; a mechanic has more; engineers still more; there are learned men, who make it their business to know all about it; and businessmen know a good deal.

A whole set of knowledge

We set down, in an easy chapter, some of the principles of it.

V

LUBRICATION AND OIL

Make it go easy

To lubricate is to facilitate slipping; oil is one of the means by which it is done.

In all machinery, rubbing of metal on metal occurs in bearings gears and slides; it is all one thing, rubbing. You call it rubbing, when hard; you call it slipping, when easy. We lubricate parts that rub, to make them slip.

Friction

All rubbing makes friction; the harder the rubbing, the greater the friction.

Friction does several things: (1) it takes power; (2) it wears the rubbing parts; (3) it heats them; (4) excess of it stops the machine—you stop it yourself, lest you get into trouble.

Adaptations of oil in the system

Hard rubbing (rubbing with pressure) is eased to slipping by a little thick oil, if the rubbing is slow; by a profusion of thinner oil, if the rubbing is fast. Light rubbing (with little pressure) is eased by thin oil. The greater the speed, the thinner and more profuse the oil. If an oil (when used) is exposed to steam (such as cylinder oil) it must be such that the steam will give the last touch to it—cylinder oil is made to be finished by steam at the instant of use.

The last paragraph gives a bare glimpse at our work; we print it, however, to state the principles.

VI

FRICTION

Everybody knows that friction

All four well known

(1) takes power
(2) wears
(3) heats
(4) makes trouble

even to the extent of stopping the work; and that is substantially all there is to know about it.

All but one forgotten

But most men, who own, or manage, or earn their living and spend their lives with, machinery, actually half-forget that their usual friction consumes a great deal of power, wears out their machinery, and costs a great deal of money.

They oil for "trouble" not for ease

They use plenty of oil, and have no "trouble." They keep down "trouble" with oil; and think nothing of power and wear. So far as oil is concerned, their minds are all taken up with "trouble," with keeping it down with oil, with the cost of oil; their minds are not on friction, on keeping that down with oil, on the value of oil.

Mistake

It takes a good deal of friction to heat a bearing enough to make "trouble," you think.

The correction of it

No; you have one hot journal, and a hundred cool ones. Before that one got hot, you saw no friction in any of them. It was in them all. How much? About $\frac{1}{10}$ hp per average bearing.* Just before that one got hot, there was almost as much friction in it as when it began to get hot. As it was the last straw that broke the camel's back; so it was only a little additional friction that made your journal begin to get hot.

Let us be as exact as we can. Suppose that the friction in the hot journal, when it was cool, before it began to get hot, was $\frac{1}{10}$ hp; the friction then doubled in that one bearing—and the journal began to get hot.

The friction that shows by heat is $\frac{1}{10}$ hp; the friction that made no sign, in that same journal, was $\frac{1}{10}$ hp. Multiply this $\frac{1}{10}$ hp by 100, because you have 100 journals. Your total friction is, roughly, 10 hp. The friction that started the "trouble" was only $\frac{1}{100}$ part of the 10 hp.

*Thurston puts it at that, on the average.

The $\frac{1}{10}$ hp scares you, because you see it. The 10 hp you are indifferent to, because you don't see it or feel it; you don't know how much it is, nor whether it may not be inevitable; you forget it. You know it exists, but regard it as you regard the weather. You see the danger not the actual harm

You are entirely right in giving instant attention to that hot bearing, because there's trouble ahead. You are wrong in not giving heed to all your bearings, because you are losing 10 hp, and because that 10 hp is wearing out your machinery. Right and wrong

There is a great deal more, than we choose to bring out, in this chapter. Let that come later. Our present purpose is only to show that you habitually overlook your losses by friction, until it begins to make "trouble;" and that the fear of "trouble" diverts your attention from loss of power, and wear. The mistake of practical men

Sometimes a physician does not distinguish his patient's disease, because his attention is only on symptoms. The disease in your works is friction; the symptom is "trouble." All your bearings have the disease; only one is in "trouble." He don't cure it You don't cure it

The cause of the "trouble" is friction. Attend to the friction. In curing that, you cure all its effects: loss of power, and wear, How to cure it as well as "trouble." Of course, you attend to the "trouble;" but, if you stop at that, your friction goes on: its other effects go on: your losses go on.

Almost all your loss, both of power and by wear, occurs when the friction is making no sign. A sudden increase of friction may cause you to stop a machine, or part of your works. It is only annoying and threatening: does no harm, because you attend to it promptly—you've got to attend to it promptly.

VII

"TROUBLE"

The cause is friction

What is this "trouble"? It is, in part, the effect of heat.

It begins from one of two causes: either the oil is not fit for the place where the "trouble" occurs, or the bearing is out of adjustment; in either case, excessive friction produces more heat than the metal conducts away; and the parts get hot, including the oil. The oil, which may or may not have been fit for the place to begin with, gets less fit, as it gets hotter; until, if not stopped, the heat brings on a crisis.

We shall not pay you so poor a compliment as to ask you to read about "trouble;" for all concerned are acquainted with it; you have the means of avoiding it, and mostly avoid it. We mention this rock above water, only to mark the two other rocks that are under water. The rocks under water are waste of power, and wear. They extend a long distance this side of the rock above water. Before a bearing gets to the stage of "trouble," that bearing is wasting much power, and wearing out machinery.

The man who attends to waste of power and wear never has any "trouble"

Pain is the self-preserving provision of Nature, in man and animals; fear of it makes us careful not to get hurt. So the dread of "trouble" is all a mill-man has, to keep him from wasting power, and wearing machinery, until he sees that his usual friction, though invisible, wastes power, and wears machinery.

Both beneficent

"Trouble" comes of a little more friction: more than you had in that bearing an hour ago. Attend to the friction you had an hour ago, and you'll never have any "trouble," unless some mechanical fault gives rise to it.

Don't wait for "trouble"

VIII

STANDARDS OF LUBRICATION

If a mill is lubricated only enough to keep its machinery going cool, and scarcely that,

Low lubrication

its standard is low; if it has a good deal of "trouble," its standard is very low; if it never has any "trouble," its standard may or may not be high, it is not very low; if the management studies economy, both of power and wear, its standard is high.

High lubrication

Usual standard low

The usual standard is: on the edge of "trouble"—not much margin. The average mill-man's notion of full lubrication is: no "trouble." He says, with great satisfaction: "We never have any trouble." Or, if he can't say that, he says: "We rarely have any trouble." The usual standard is low.

The high standard

The high standard is: to have no "trouble," of course, but not stop there: to economize power and wear of machinery.

Friction invisible

All the difficulty of this subject comes of these two facts: (1) the "trouble" compels attention; (2) mere friction does not. Some loss of power and wear are inevitable; and few know whether their losses of power and wear are excessive or not, unless there is "trouble."

Embarrassing

The difficulty is ours as well as yours. We make the oils to save your power and machinery; but, when we offer them to you, how can we show you whether they do these things or not?

There is an easier difficulty on top of this. Let us take one minute to get rid of it.

IX

MISTAKING THE COST OF OIL

You pay for oil by the gallon: so much a gallon. That is the handiest way to think of the cost of it: so much a gallon. There is another way: by its lasting: so much an hour, or month, or year.

By the gallon or year

If you think of the cost by the gallon, you want a low price by the gallon; but, if you think by the hour or month or year, you want a low cost by the hour or month or year.

Two different things

If you think by the gallon, the lower the price, the better; if by the hour or month or year, the lower the cost by the hour or month or year, the better.

The two do not go together

Which do you want: low price, or low cost? Which makes the less invoice? cheque? account?

Only one is right

You want the oil that costs least by the hour and month and year. One gallon is as good as ten, if it lasts as long.

By time

But the custom is, with many: to take the low-price-by-the-gallon oil! They think one

One gallon is not as good as another

gallon as good as another, and take the low-price gallon.

Queer

This reckoning price as cost is a queer mistake for a businessman; but, so many make it, there must be a reason for it.

Less by the gallon more by the year

Here is the difficulty. They don't believe that the oil that costs less by the gallon, costs more by the hour and year; but it does.*

They don't buy boots that way—they know leather too well for that. They buy boots of a sort to answer the purpose of boots.

That is: men buy what they know, with their knowledge; what they are not acquainted with, they buy with their not-knowledge. We, all of us, have too much of that sort of knowledge.

The price has nothing to do with it

The price of oil by the gallon is of no importance at all; even its cost by the year is of little importance. Keep your mind on what oil is for: on ease: mechanical ease.

The oil that keeps friction down, is the oil that makes dividends.

* Of course, the prices of oils are not fixed by the time they last; but the general fact is that high-price oils are durable; low-price not.

X

THE LOSSES BY FRICTION THAT MAKE NO SIGN

We cannot go through your works, put our hand on the friction-places, one after another, and tell you how much you are losing in each by preventable friction—how much in power, how much in wear. That is impossible.

Our knowledge of friction is not a knowledge of your conditions

There are ways of measuring losses of power, not always practicable; and it is not always worth while to do it, when practicable—the business may not be enough to warrant it.

It is knowledge nevertheless

In most works, the manager has his choice of educating himself out of wrong opinions about lubrication, into the settled principles of it; then making himself acquainted with the experiences of others, and being guided intelligently by them—or he may go on as he is.

Two ways for a manager

In the main, we must be content with the business of those who will study the subject a little.

Only one for us

XI

WHAT BECOMES OF LOST POWER

No power is lost

When something is burnt, consumed, worn out, dried up, decayed—destroyed, as we say—it is all left: it is only changed. As, if stolen, somebody else has it. Not even force is lost: no power is ever lost.

It goes into wear of machinery

What becomes of the power you "lose" by friction? There is no mystery in it: every horse-power of it is consumed in wearing the parts that rub; precisely the same as your power not "lost" is consumed in doing your work. The very expression used in stating its quantity, "horse-power," shows it is still power, though lost to your service and turned to your injury.

Useful power

Injurious power

What friction leaves is useful power, the power that does your work; what friction takes is wear and tear, a part of which is inevitable; what preventable friction takes is waste, by ignorance, negligence, choice, good or bad judgment—small waste is not worth the attention required to stop it.

Friction heat

We say, a few lines above, the power you lose is "consumed"; and, in the first line of this chapter, "consumed" means changed.

What change is that? The power you lose by friction is, all of it, turned to heat.*

Conducted away by the metal

This accounts for hot journals; but why are not all journals hot—there is friction at all? There is heat at all journals, in the exact proportion of friction: 42.42 thermal units per hp lost. The reason some journals "heat," and others do not, is that heat is conducted away from most journals about as fast as it is developed, but not in hard ones. The difference, between one hard journal and all the rest, may not be in *production* of heat, but in *conduction* of it.

Look out for unsuspected friction

Here is another view of the now familiar fact that your friction enemy is, not at your one hard journal, but at every friction-place in your works.

XII

ANOTHER PRINCIPLE

We must add to Chapter V what would have been uninteresting, if we had asked you to read it there.

*At the rate of 42.42 thermal units per hp lost. A thermal unit is the amount of heat that raises the temperature of one pound of water one degree Fahrenheit.

Law of friction with metals

The friction between an unlubricated journal and bearing varies with pressure: double the pressure doubles the friction; treble the pressure trebles the friction; ten times the pressure makes ten times the friction; and so on.

Law of friction in oil

The friction between the parts of a liquid or fluid is not increased by pressure at all. For instances: water and air are equally liquid and fluid at all pressures; so oil is equally liquid at all pressures. And friction in water or air or oil is not increased by pressure. Say nothing of speed at present.

Almost true

Now: full lubrication means that the friction is all taken off from the metal and left in the oil. Taken off from the metal, where pressure affects it; and put in the oil, where pressure does not affect it.

Entirely true

The friction between a lubricated journal and bearing is, so far as those surfaces touch one another, affected by pressure; so far as they are kept apart by the oil, the friction is not affected by pressure.*

Full lubrication, like any other perfection, is not attainable. High lubrication means about as near an approach to full as is practicable.

* Where mixed friction (friction of metals and friction of oil) is met with, its laws approximate those of solid friction as the journal is run dry, and those of fluid friction as it is flooded with oil.—*Thurston*.

High lubrication, therefore, is less affected by pressure; scant lubrication is more affected by pressure. Easier language to say the same thing: A poor oil is very poor indeed in a hard place; a middling oil is poor in a hard place; the very best is none too good in a hard place.

This also

This explains that "trouble," and takes all mystery out of it. That particular bearing, whatever it is, is not half lubricated. It may be full of oil; no matter; the oil is not right for the place.*

If the "trouble" is not mechanical

The principle is: lubrication takes friction away from hard metal, and puts it in easy oil. But the oil must suit the place, as appears in the following chapters.

The principle

* Of course there may be a mechanical fault in the bearing: out of adjustment or bad design.

Here is a good place to tell of an incident at the Electrical Exposition in Philadelphia, 1884. Nearly all the exhibits were oiled with Vacuum oil; and our principal part in the Exposition was this card on the exhibits: "This exhibit is lubricated with Vacuum oil."

One exhibit, not using Vacuum oil, was obliged to stop on account of a hot journal. With Vacuum oil, it began to cool immediately. Another, though using Vacuum oil, was obliged to stop in twenty minutes. This was an enormous dynamo, which turned out impracticable: could not be lubricated.

XIII

FRICTION IN OIL ITSELF

A simple way of putting it

Let us look a little closer at it. A part of the oil, between journal and bearing, sticks to the journal and turns; and a part of it sticks to the bearing and does not turn. These two thin films of oil (the journal film and the bearing film) slip upon and along one another. There is some friction between the two films of oil. If we have got journal and bearing apart by putting oil between them; and kept them apart by using an oil that is stiff enough to do that very thing; we have full lubrication, so far as journal and bearing go.

Viscosity

Now consider the friction between the two films of oil: the journal film and the bearing film. The stiffer the oil, the more there is of this interior friction in it. The thinner the oil, the less there is of this interior friction in it. Full lubrication is got by the thinnest oil that keeps journal and bearing apart.

Thick and thin oil

Thick-enough oil, to keep journal and bearing apart, appears to have a good deal of this friction inside of it. Thin-enough oil, to make the least possible friction inside of the oil, appears to be too weak to keep journal and bearing apart.

These appearances are misleading. The actual friction in oil itself is very light indeed at slow and moderate speeds, and pressure does not affect it; and very thin oil keeps journal and bearing apart at high speeds—speed carries the oil in between them.

The facts are not as they seem

The friction in oil itself depends on the speed.

XIV

THE CONDITIONS FOR OIL

The conditions for oil are pressure and speed. Besides, there are limits of temperature; for oil undergoes a change, with heat and cold: it is thinned by heat and thickened by cold.

Pressure speed and temperature

You do not know, and cannot find out, what the pressure is at a bearing. It depends on the accuracy of adjustment. It differs at different parts of the bearing.

The pressure at any point is always uncertain

XV

EFFECT OF PRESSURE

It tends to force the lubricant out from between the metals in contact, and leave them dry; which, of course, is fatal. Thick oil resists this tendency, more than thin, at slow and moderate speeds.

Pushes oil out and prevents lubrication

Otherwise no effect

But pressure has no effect on the friction of oil upon oil.

XVI

EFFECTS OF SPEED

Carries oil in and supports lubrication

The slower the speed, the stiffer the oil must be to keep journal and bearing apart; and the higher the speed, the thinner the oil *may* be, and still keep journal and bearing apart. The reason is: speed takes the oil in with a certain force.

Increases the friction of oil upon oil enormously

The friction of oil upon oil is governed by speed, and is not much affected by anything else.*

XVII

A COMPROMISE

The two parts of friction

The friction at any journal is, of course, the sum of the two elements: rubbing be-

* The law of friction in oil itself (or in any other liquid or fluid) at high speeds, is: the friction goes up with the square of the speed. Thus: 100 times a moderate speed makes 10,000 times the friction in oil itself.

It is clear that this interior friction in heavy oil, at low or moderate speed, is less than one would think; and, at high speed, is almost incredibly large. The latter accounts for the largest losses of power, as shown by the diagram at the end of this book.

tween the two metals, and rubbing between the two films of oil.*

The least possible sum of the two is full lubrication

At all pressures and speeds, we get the least possible friction with oil that is heavy enough to keep journal and bearing apart, and yet light enough to avoid excessive friction inside of the oil itself. Of course, we never get to a perfect compromise—that would be full lubrication.

XVIII

HOW MUCH DOES LOSS OF POWER AMOUNT TO

Only the wear

If you use water-power, and have it to spare, your loss of power is nothing but wear. If you have a mill in the lumber regions, burn waste wood, and are glad to get rid of it, loss of power is a gain, except the wear that accompanies it.

Only the wear

If you make steam, and choose to make more than you need, your loss of power, perhaps, is nothing but wear.

* Thurston says: "At a pressure sufficient to force out "the lubricant, or to prevent its entrance between the "metal surfaces, we have solid friction; when the pressure "is light, and the lubricant has sufficient capillarity to "separate the two metals, we have fluid friction. Between "these two conditions, we have both solid and fluid fric-"tion. Low speed favors the former, and high speed the "latter."

The wear and the cost of lost power

If you make steam economically, your loss of power amounts to whatever it costs to produce the power you lose, and the wear.

Only the wear

If you buy your power, and choose to buy an excess, your loss of power, perhaps, may be nothing but wear.

The wear and the cost of lost power

If you buy economically, your loss is whatever you pay for the power you lose, and the wear.

Inseparable

You cannot waste power alone: it takes machinery with it.

XIX

HOW MUCH DOES WEAR AMOUNT TO

Bill of items

Whatever preventable friction you have, at a journal, is wearing that journal and bearing. This wear is a loss of the cost of repairs; of the time of repairs (and stopping is inconvenient—that is loss); and the journal and bearing work less well, and must be replaced by new at an earlier date, because of this wear by preventable friction.

The items vary

These losses occur at every friction-place in your works. There is no way to state them in detail and generally. Professor Thurston speaks of 2½ per cent of the plant as a fair annual allowance for putting in new machinery, solely for wear. Repairs are perhaps too variable for estimate; because works generally are oiled with one eye on the price of the oil and the other on "trouble;" and friction and wear are not much thought of. Generally, as the degree of lubrication is an accident, so is the wear.

XX

TOTAL GAIN

We have seen that high lubrication avoids the usual wastes of preventable friction. These are of power and machinery. Look at these savings separately. Good management

In saving power: you burn less coal, have less to handle; your works are under less strain, have fewer break-downs and accidents, lose less time (of works and staff and men); and your business goes on more regularly. What goes with power

In saving wear: you have less repairs, less expense for repairs, less effect of wear and repairs in deterioration of plant—for repairs do not make machinery new again—you have longer life of machinery, and less interruption of work. What goes with wear

Every one of these gains is actual, whether or no ascertainable; which they are not. Actual gains

And the sum of them all is as clear as if it were ascertainable; which it is not.

XXI

PERFECTLY KNOWN TO SCHOLARS

Nothing new

Engineers—we mean mathematicians who apply themselves to mechanics—know all about it, and wonder that practical men, those most concerned, should doubt these commonplaces of science.

XXII

PRACTICAL MEN ARE FINDING IT OUT

The law of gravity ruled before Sir Isaac Newton found it out

But it is not strange that a man, who thinks of oil by its price per gallon, and does not consider how long it lasts, should also not see what it does, or know whether it does it; that he should imagine that anything called a cylinder oil is fit to put into a cylinder; anything called an engine oil is fit to put on an engine; and so all through. To him, one oil is as good as another, so long as he sees no friction. There is no friction, for him, when journals run cool. Lubrication is nonsense, to him, except as machinery cries

The law of friction rules whether you regard it or not

for oil, or actually gets hot and has to be stopped.

But the Experience chapter shows that practical men are finding out their mistake; and our business shows it.

XXIII

"WE NEVER HAVE ANY TROUBLE"

Now, if we put together what we have learned about lubrication and oil—a part of the a b c of the subject—we see that hot journals have almost nothing to do with it.

"No trouble" is no assurance of high lubrication

The friction, that causes hot journals, has unsuspected preventable friction behind it; *which wastes more power, and wears the machinery vastly more, than the friction that shows in heat. Because this unsuspected friction is working all the time and all through the mill;* while the friction that shows itself by heat is generally asleep, and never wakes up except at one or two journals.

Look out for insensible friction all through the mill

This is not true of all works. There are works, in which lubrication has been pursued about as well, and as fast, and as far, as it can be pursued; but the manager never says: "We have no trouble."

Power-and-wear economy far above "no trouble"

XXIV

EXPERIENCE SHOWS

Is the doctrine practical

Here is the place for experiences; but they make a long chapter; you may prefer to apply, to Vacuum oil, the doctrine set forth in the foregoing discussion.

We put the experiences at the end of the book, in a chapter by themselves. Turn to that, if you like—page 44—but do not lose the connection here, if you feel disposed to consider our product.

XXV

FRICTION CAN BE REDUCED UNLESS IT HAS BEEN INTELLIGENTLY AND DILIGENTLY PURSUED

It is a problem to solve not a riddle to guess

The manager, whether able or not to find out that a substantial saving of friction is practicable in his particular works, should conclude that it can be made, unless he has already made it.

Vacuum oil is the means

The way to do it is with Vacuum oil—the only way to do it is with Vacuum oil. If he thinks he has done it otherwise, it is still to be done with Vacuum oil. This is always and everywhere true, but it cannot always be shown.

Do you ask us to show you, in your works, that you can do it? Our business is to supply the oils to do it; and, where the trade is enough to justify the expense, we are always ready to send a man to show how to do it there. We're doing it all the time all over the world

XXVI

ENDURANCE OF VACUUM OIL

We have scarcely mentioned the wear of oil: how long it lasts: how long a gallon lasts. We barely alluded to it in treating of cost, in Chapter IX, page 17. Durability

The general rule is: a poor oil is used profusely, and is soon gone; a good one is sparingly used, and lasts from two to twenty times as long—or would, if fed economically. The general rule

But the rule is not invariable. For instance: in spindle oil, the virtue of being extremely light (to save power) is more important than lasting long (to save cost); and the two do not go together. An exception

In some high-speed machinery, oils are used profusely and filtered and used again Filtering

and again.* The endurance of oils, to be filtered, depends on the number of times they can be filtered and used.

Particulars

The Vacuum high-speed-machinery oils can be filtered and used—we do not know how many times—indefinitely. Our cylinder oils and general oils last several times as long as the common oils. Velocité spindle oil is an exception: it wears about the same as other light spindle oils.

* Whatever Vacuum oil you can catch you can filter profitably, even cylinder oil. This is less true of inferior oils.

A good filter can be made by laying a fold of cheesecloth over three inches of white cotton waste well picked, this on fine felting about a half-inch thick, the whole on a perforated sheet, or fine wire cloth. The oil, after passing through the above, drops into a funnel, at the bottom of which is a half-inch pipe, at lower end of which is a funnel-shape flange of a smaller diameter than funnel at top end of pipe. The small pipe is of such length that lower funnel comes within one and one-quarter inches of bottom of tank or vessel, the tank or vessel having at least 14 inches of water.

This kind of filter is very easy to make, and easy to clean. The working of it is as follows: The cheesecloth gathers all large dirt which can be shaken off. The cotton waste takes the next in size, and the felt the rest; after which, the oil passes through the water, and all foreign matter remaining in the oil, is taken out.

You can filter anywhere out of dust; the warmer the better.

If you filter much oil, you may prefer to buy a filter with a steam coil in it. It filters faster, not better.

XXVII

COMPARATIVE COST OF VACUUM OIL

Vacuum oils are sold at higher prices than other oils; they have to be; they cost more to make. As they last (with one exception) from two to twenty times as long, it is fair to say that they cost less by the month and year. We suppose they cost about half as much as others, in actual use. They would cost, perhaps, not more than a quarter as much, if used economically. *Perhaps half on the average*

A long-lasting oil is not likely to last as long as it would if not wasted; the better the oil, the more it is wasted. An oil, that would go ten times as far as a usual oil, gets fed in a way to go perhaps three or four times as far. *Correct*

XXVIII

COMPARATIVE VALUE OF VACUUM OIL

The difference, in value, between two oils—and between two sets of oils—is the cost of what one saves and makes, and the other does not. *A hundred to one sometimes*

Good oil saves coal, repairs, detentions, break-downs, life of machinery; makes increased efficiency, larger and better results, regularity, comfort, credit. *Particulars*

Accuracy impossible

Some of these things can be measured more or less accurately; some cannot be measured at all. They all count for profit or loss: they all have a money equivalent.

The world is full of poor oil

Poor oil does the same things to a less extent. The difference cannot be measured; but some of the parts of it can sometimes.

We presume it was worth a hundred to one

An English railway experience, told in detail on page 44, shows that Vacuum cylinder oil was worth $4.83 (19s 11d) a gallon more than the Company's cylinder oil, in coal alone. There was no refacing of valves for a year—no need of it—three repairs and detentions saved. The railway had 557 engines: it had too many, with Vacuum oil, or could do more work; and those engines had longer life, with Vacuum oil. That oil was worth, altogether, more than its value for saving coal alone: it had another value for saving repairs: it had still another for saving investment in plant. Only one of these values was measured in money, or could be measured in money.*

* If we could get at the rate of deterioration of any valuable machinery, driven at anywhere near its limit of power, we should have a new measure of the value of oil. The value of oil that reduces wear, in any such ratio as that revealed in this English experience, must be several-hundred dollars a year, *in terms of wear alone*, for one locomotive.

The foot-note, page 42, shows Vacuum spindle oil worth more, than the spindle oil of the mill, by $1600 (£330) a year for the spinning-room; how much per gallon we cannot compute, not having the data.

Twenty to one very likely

The railway saved half of its previous yearly expense for oil; the cotton-mill spent for oil, we suppose, about the same as before.

Minor incidents

Some experiences show gain in one way, and some in another. There are more gains than appear; we cannot see them all, and can rarely count any of them accurately.

There are more gains than appear

Evidently, the comparative value of Vacuum oil is very large.* It costs less by a half or two-thirds or three-quarters; which is of little account. It saves and makes more—ten twenty thirty times more—and that is what oil is for.

Ten to one is enough

* You value an oil, that is offered to you, in terms of the oil you are using; which is right, and easy to reckon. Nobody cares how much more an oil is worth than no oil at all.

The question is what shall I gain or lose by using a different oil? You will find the answer surprising and interesting. We may as well quote Professor Thurston on it.

"The consumer could, in such a case, better afford to pay "$5, or perhaps even more, per gallon for the good oil, "than accept the less valuable lubricant as a gift."

Again: "A reduction of 50 per cent, in the work lost by "friction, has often been secured by a change in the lubri-"cant." And he says that 50 per cent is a fair estimate of the average loss in a mill by friction.

XXIX

RATE OF FEED OF VACUUM OIL

Slow

Our cylinder oil is fed very slowly, compared with the usual rates of other cylinder oils. A drop a minute is as slow as most lubricators can be depended on to feed, without danger of getting obstructed; and that is enough, of 600 W, for the average 100-hp engine.

Too fast

The usual practice is to feed too fast by several times—it is not considered worth while to watch the lubricator closely enough to feed no more than is needed—and yet the usual feed is perhaps a quarter as much as of other cylinder oils.

Slow

Slow-speed-machinery oils are fed, in all modern machinery, by self-acting devices that can be adjusted to any rate of supply; and the rate of supply is generally, with Vacuum oils, about half or a third or a quarter as much as with other slow-speed-machinery oils.

Fast and filter

High-speed-machinery oils are fed in a continuous stream—the Vacuum oils as fast as any, no matter how fast—and filtered. The difference, as to endurance, shows in the number of times they are used. The

Vacuum oils are filtered indefinitely—there appears to be no limit. And the quality is, so far as we know, very little affected by use.

XXX

METHOD OF FEED

Professor Thurston says "The method of "supply is a matter of supreme importance."

Hand oiling extravagant

The old way was by hand. That primitive method still prevails, in rough shops and old machinery. There is a great variety of self-acting oil-feeding appliances, some of them excellent, made to put on this old machinery.

Automatic appliances

There is only one good way to oil any part of any machine: automatically. Hand oiling is every-way wasteful and unsatisfactory: it alternately floods and leaves the machinery dry: it throws most of the oil away. Machinery needs, not alternate drenching and drought; but a uniform flow, enough and no more. There is as much difference between a well- and ill-oiled machine, as between a good and a bad one; indeed a good one becomes a bad one, not only when ill-oiled, but by being ill-oiled. No machinery ever arrives at good working condition without such regular oiling.

Self-oiling machinery

Modern machinery is self-oiling. It makes this chapter superfluous, where it is used. Those appliances make old-fashion machinery self-oiling.

Professor Thurston

Thurston says "The oil-bath sometimes "reduces friction to one-tenth the amount "observed with the more common methods... "An oil-pad, as in railway practice, is better "than an oil-cup; and a bath is much better "than either." Again: "The art of economi-"cal employment of lubricants consists mainly "in the determination of their adaptation to "specific purposes, and in the application to "each machine in which pressures on lubri-"cated surfaces of widely differing amounts "are found—of precisely that quality of un-"guent which is best adapted to that par-"ticular place; and, *above all*, applying it in "the best possible way."

Constant uniform adequate quantity

Our purpose, here, is to show the importance of oiling with constant uniform adequate quantity: just enough oil to make friction as little as possible. Elsewhere we treat of quality, suitableness of the oil to its use; here only of quantity, neither to sell more oil, nor save waste, but to show the value of oil beyond your belief, and to show the means of getting that value: feed automatically.

Self-oiling appliances pay for themselves, in a few weeks, in power and wear and oil—the last is the least of the three; but that is the one you can measure. Cost nothing

If it pays to use a good-enough oil to do more than keep out of "trouble," to save more power and wear; if a constant uniform flow of that oil still more saves power and wear, to the extent of nine-tenths, as Thurston says in the case of the bath against common methods; it pays, "above all," to secure that constant uniform flow. and pay

We do not know your works, nor therefore the feeding-devices that suit your particular needs. We supply the general knowledge; it is your part to apply it. We should be glad to do that too, if we could. We mean: we cannot set the particulars down in a book. Our salesmen do a good deal in the way of advising, and even supplying, appliances. We supply automatic appliances

XXXI

GREASE

The principles of lubrication with grease are precisely the same as with oil; for grease, Same as oil when actually in use, is melted; and melted grease is oil.

Preferred for some uses

Grease is preferred to oil, in some circumstances, notwithstanding the loss of power it involves.

We make greases also.

XXXII

WHERE VACUUM OIL IS SOLD

Our oil is sold in every country where there is machinery.

XXXIII

OUR WORK CONTRASTED WITH THAT OF OTHERS

Our work for thirty years

We have been trying, for thirty years, to create such a system of oils as this book explains the importance of. And we seem to have accomplished a part of that great undertaking; which is no less than offering smoothness ease and economy to the mechanical work of the whole world.*

* The saving of power just now, by putting our spindle oil in a cotton-mill was .42 hp per frame. The owner computes the saving in money at $1600 (£330) a year for the spinning-room.

We long since solved the problem of cylinder oil, in the judgments of both scientific and practical men all over the world. We may say the same of oil for high-speed engines and dynamos; nearly the same of a general engine oil; and measurably of spindle oil.

Examples

The other "oil-manufacturers" do not take this serious view of their business. They mostly mix—there are few manufacturers—their object is only to sell; and their oils are as good as could be expected under the circumstances.

The work of others

We are able to raise our standard yearly. It is, of course, a rare achievement to add an oil to our list; but our work is improving all the time, from new knowledge skill and facilities. The standards of others appear to remain exactly the same from decade to decade: they make to sell, and only to sell. Competition, however, forces some improvement of oils upon them.

Improving all the time

Improving a little

XXXIV

NATURE A BETTER ALLY THAN ENEMY

In a book on reducing friction with oil, we can only tell what the laws of nature are, on friction and oil, how men make money by going

It is better to go with the laws of nature

with those laws, and lose it by going against them. We have stated the laws and explained them as well as we can, pointing out the mistakes men commonly make. It remains to confirm the whole by quoting experiences of practical men.

XXXV

EXPERIENCES

Common experiences

Experiences abound. They come to us in business letters; are generally expressions of good-will for services rendered. Few of them show the differences in oils that count for profit or loss in the larger ways.

Tests

One of the most satisfactory sources of knowledge, of any oil, is its work in a mill—no matter what sort of a mill—alongside of another oil in the same conditions. No; not alongside of it; after it. First one oil, and then the other: both in the same conditions, as near as they can be the same at different times.

Best so far as it goes

This is a "test." It is made with the greatest exactness and carefulness; but tests are not made every day, or in every mill; and they cannot be made to answer all the questions, which both the reader and we should like to have answered.

One of the most interesting is that of an English railway—printed reference not permitted. Our salesman prevailed on the proper officer to try our cylinder oil on one locomotive. The officer said it would come to nothing; for he was satisfied with his oil (which cost 10d), and "would not pay 3s for any oil." He agreed to try it, however, on one locomotive.

Distinguished example

There was no perceptible difference, except that less oil was used, and seven to ten cwt less coal was burned on a trip.

Two differences

He was not surprised on finding less oil required; but so much less coal was very surprising. He put the oil on another locomotive with like result; and then on ten others, still with the same result.

One of them was surprising

Could the change of cylinder oil account for burning so much less coal? He would find out. Went back to his usual oil—and burned his usual coal. Went to Vacuum oil again, and had coal left over, same as before.

Tested the test

It was proof.

It was proof

We have seen the "performance sheets" of those twelve locomotives that year, and of the preceding year; and, from them, computed the saving per year, in the entire locomotive service, on both oil and coal. On oil, it was nearly £500 ($2500); on

$2500 (£500) in oil

$15,000 (£3000) in coal

coal it was over £3000 ($15,000). The year's supply of cylinder oil for 557 locomotives was, roughly, three-thousand gallons; and every gallon of Vacuum cylinder oil, at three shillings a gallon, saved eight gallons of ten-penny oil, three tons of coal at six shillings a ton, and all repairs.

No sensible wear in a year

It had been the practice to reface the valves about once in four months. At the end of a year, with Vacuum oil, the valves of those twelve locomotives had not been refaced; and they did not need refacing—the tool-marks on them were fresh.

Bates Manufacturing Co's cotton-mills, at Lewiston, Maine, U S A, are driven by water-power. One of the water-wheels drives 8000 warp spindles and all the pickers in one mill. This wheel was always hard-pushed with former oils: a good chance to test oil.

Velocité spindle oil was tried in that mill. The result was an instant gain in speed, and, in two days, there was two-inches spare gate on the wheel.

Then the previous oil was put on the spindles; and the wheel went back to its usual speed.

Darwen Spinning Co Limited, Blackburn, England, saved 13.76 hp by using Vacuum cylinder oil on an over-worked 500-hp engine. The saving of coal was one or two tons a week; the saving in cost of oil was about one-half.

Tangyes Limited, Engineers, Cornwall Works, Birmingham, England.—"We have used your "cylinder engine and machinery oils for twelve "years. Comparative tests have shown that a "less quantity gives more perfect lubrication than "of other lubricants. Your oils have caused no "corrosion gumming sediment or other bad "result. We recommend them to purchasers of "our engines and machinery."

Duckworth Howarth & Co (Victoria Mill), near Blackburn, England.—A most instructive example.

25 August	1891	indicated hp 207
17 December	1891	indicated hp 200
24 June	1892	indicated hp 176

On the first date, 600 W cylinder, Vacuoline engine, Vacuum shafting and loom, oils were put on. In four months the saving of power was 3½ per cent; in ten months 15 per cent. Meantime no change had been made in engine machinery or load, beyond that caused by the oils.

The immediate saving was due, of course, to improved lubrication alone; the ultimate saving was due to the easier working, caused by the *improved condition of the bearing surfaces*—polish instead of roughness—*which always comes upon bearings in wholesome wear with high lubrication.*

There is an instant saving of power, on changing from bad to good oil, from good to better, or better to best; but the immediate change is only part of the gain. Perhaps the gain that comes by this polishing process, though gradual and likely to be overlooked because it is so gradual, is always the principal part of the gain when the mechanical superintendence is intelligent skilful and careful.

Birkin Brothers, Lace Manufacturers, Nottingham, England.—"Lubrication by your 600 W "cylinder oil is so much more perfect than that "by any of the various lubricants we have used, "that we are saving something like twelve loads "of coal in six months, when doing the same "work or more."

Stanton Iron Works Limited, near Nottingham, England.—"600 W and Red Z save power and give satisfaction every way."

Derwent Foundry, Derby, England.—"We are "highly pleased with 600 W in every way. It "effects a decided saving in coal."

Milwaukee Worsted Mills, Milwaukee, Wisconsin, U S A.—Running without load, the indicated hp was 52.10 and 59.75 respectively with 600 W and Vacuoline, and with other oils: 7.65 hp difference.

Willer Mfg Co, Milwaukee, Wisconsin, U S A. —A fall of six in the indicated hp came with using 600 W; and, in the cost of oil, from 12 to 3 cents a day.

Eckhart & Swan, Chicago, Illinois, U S A.—Flour-Mill tested 600 W and the cylinder oil of the mill in twelve-hours work with this result:

	with 600 W	with other oil	difference
Flour ground	281 bbl	275 bbl	6 bbl
Coal burned	9,933 lb	10,140	207 lb
Friction	120.41 hp	132.76 hp	12.29 hp
Cost of oil	4 cents	9 cents	5 cents

Société Anonyme "La Dendre," Termonde, Belgium.—With former oil, were using 48.40 per cent more power for the same work.

MM Omer Vanham & Cie, Braine-le-Chateau, Belgium.—With former oil, were using 47.54 per cent more power for the same work.

Goosery Cotton-Mill, Calcutta, India, used a native spindle oil two months, then Vacuum spindle oil seven months, with these averages:

	Mds of coal	Turns of engine per month	Indicated hp
Vacuum oil . . .	9,485	2,333,462	873
Native oil	12,808	2,269,245	1189
*Vacuum oil . . .	9,224	2,269,245	849
Difference . . .	3,584		340

With former oil, they were using 40 per cent more power for the same work, and burning 38.84 per cent more coal.

MM Baden & Fils, Barentin (Seine-Inférieure) France.—With former oil, were using 39.17 per cent more power for the same work.

* This line is the second line above it reduced to an even number of turns of engine with that of the native oil.

MM Waddington Fils & Cie, Saint-Rémy-sur-Avre (Eure et Loire) France.—With former oil, were using 38.98 per cent more power for the same work.

Spett Cotonificio Bergamasco, Ponte di Nossa (Valseriana), Italy.—With former oil, were using 38.46 per cent more power for the same work.

MM Leon Crèpy & Fils, à Canteleu-les-Lille (Nord), France.—With former oil, were using 38.43 per cent more power for the same work.

Companhia de Fiação de Crestuma, Crestuma, Porto, Portugal.—With former oil, were using 35.17 per cent more power for the same work.

Fukushima Boseki Kabushiki Kwaisha, Osaka, Japan.—With former oil, were using 34.85 per cent more power for the same work.

Companhia Fabril de Salgueiros, Porto, Portugal.—With former oil, were using 32.85 per cent more power for the same work.

Spinnerei Forchheim, G m b H, Forchheim, Bavaria.—With former oil, were using 29.67 per cent more power for the some work.

Ise Boseki Kabushiki Kaisha, Yokkaichi, Ise, Japan.—With former oil, were using 29.64 per cent more power for the same work.

Signori Grüber & Cia, Serravalle, Scrivia, Italy.—With former oil, were using 29.60 per cent more power for the same work.

Sobrinos de Juan Batllo, Barcelona, Spain.—With former oil, were using 28.68 per cent more power for the same work.

Westfälische Jutespinnerei & Weberei, Ahaus, Germany.—With former oil, were using 28.61 per cent more power for the same work.

Herren Gebr Rosenthal, Rankweil, Austria.—With former oil, were using 27.13 per cent more power for the same work.

Messrs Dobson & Barlow, Kay Street Works, Bolton, England, made three tests in experimental room, showing saving of 12 per cent, 26 per cent and 8½ per cent in power.

Fukuyama Boseki Kabushiki Kaisha, Fukuyama, Japan.—With former oil, were using 23.93 per cent more power for the same work.

Miye Boseki Kabushiki Kaisha, Yokkaichi, Ise, Japan.—With former oil, were using 23.58 per cent more power for the same work.

Deutsche Jutespinnerei & Weberei, Meissen, Germany.—With former oil, were using 23.25 per cent more power for the same work.

One of the largest Kammgarn-Spinnereien in Leipzig, Germany.—With former oil, were using 23.08 per cent more power for the same work.

Rosenlunds Spinnerei Aktiebolag, Göteborg, Sweden.—With former oil, were using 22.65 per cent more power for the same work.

Herren Meyer Kauffmann, G m b H, Breslau, Germany.—With former oil, were using 22.13 per cent more power for the same work.

Flli Fumagalli di G B, Peregallo, Italy.—With former oil, were using 22.05 per cent more power for the same work.

A Cotton-Mill in Hyde, England.—With former oil, were using 21.50 per cent more power for the same work.

A Cotton-Mill near Manchester, England.—With former oil, were using 21.24 per cent more power for the same work.

Asahi Cotton-Mills, Osaka, Japan.—With former oil, were using 21.23 per cent more power for the same work.

Spelty Keller & Co, Sondrio, Italy.—With former oil, were using 21.20 per cent more power for the same work.

Kurume Boseki Kabushiki Kaisha, Kurume, Japan.—With former oil, were using 20.70 per cent more power for the same work.

Signor Bernardo Caviglia, Voltri, Italy.—With former oil, were using 20.21 per cent more power for the same work.

Chinese New Cotton-Spinning Co, Shanghai, China.—With former oil, were using 19.33 per cent more power for the same work.

MM P & H Leurent, Turcoing (Nord), France.—With former oil, were using 19.26 per cent more power for the same work.

MM Normant Frères, Romorantin, France.—With former oil, were using 18.50 per cent more power for the same work.

Herr I B Limburger, Jun, Ketten (Reichenberg), Austria.—With former oil, were using 18.85 per cent more power for the same work.

A Sedo y Ca, Esparraguera, Spain.—With former oil, were using 18.84 per cent more power for the same work.

Signor Conte Raggio, Novi Ligure, Italy.—With former oil, were using 18.79 per cent more power for the same work.

Signori R Butti & Cia, Legnano, Italy.—With former oil, were using 18.18 per cent more power for the same work.

A Cotton-Mill in Preston, England.—With former oil, were using 18.16 per cent more power for the same work.

Companhia Fabril do Cavado, Ruaes (Braga), Portugal.—With former oil, were using 18 per cent more power for the same work.

Kurashiki Boseki Kabushiki Kaisha, Kurashiki, Japan.—With former oil, were using 18 per cent more power for the same work.

Companhia de Fiação e Tecidos Lisbonense, Lisboa, Portugal.—With former oil, were using 17.82 per cent more power for the same work.

Herren Jenny & Schindler, Kennelbach bei Bregenz (Vorarlberg), Austria.—With former oil, were using 17.80 per cent more power for the same work.

Jutespinnerei & Weberei Bremen, Bremen, Germany.—With former oil, were using 17.53 per cent more power for the same work.

Herren G van Delden & Co, Gronau, Wesphalia.—With former oil, were using 17.52 per cent more power for the same work.

Signori F Borghi & Cia, Legnano, Italy.—With former oil, were using 17.42 per cent more power for the same work.

Spinnerei von Carl Schmölder & Co, Rheydt, Germany.—With former oil, were using 17.39 per cent more power for the same work.

Shimmachi Bosekisho, Shimmachi, Japan.—With former oil, were using 17.19 per cent more power for the same work.

Nebsky Thread Mfg Co, St Petersburg, Russia.—With former oil, were using 17.19 per cent more power for the same work.

Filature Saint-SévER, Rouen, France.—With former oil, were using 17.12 per cent more power for the same work.

Sres Sert Hermanos e Hijos, St Andres, Barcelona, Spain.—With former oil, were using 16.67 per cent more power for the same work.

Kamgarnspinnerei Eitorf, Eitorf, Germany.—With former oil, were using 16.66 per cent more power for the same work.

Herren G Borckenstein & Sohn, Neudan, Austria.—With former oil, were using 16.29 per cent more power for the same work.

Indian Mfg Co's, Mahalukshmi, Bombay.—With former oil, were using 16 per cent more power for the same work.

Filature de Tubize, Belgium.—With former oil, were using 16 per cent more power for the same work.

Signori Fratelli Gibelli, Campoligure, Italy.—With former oil, were using 15.66 per cent more power for the same work.

Herren Pollack's Söhne, Braunau, Austria.—With former oil, were using 15.27 per cent more power for the same work.

Signori Biffi & Sironi, Canonica Lambro, Italy.—With former oil, were using 15.19 per cent more power for the same work.

Filature de Graville, Havre (Seine-Inférieure), France.—With former oil, were using 15.16 per cent more power for the same work.

Cotonificio Cantoni, Castellanza, Italy.—With former oil, were using 14.86 per cent more power for the same work.

Trombini & Cia, Milano, Italy.—With former oil, were using 15.08 per cent more power for the same work.

Spett Linificio & Canapificio Nazionale, Fara d'Adda, Italy.—With former oil, were using 14.62 per cent more power for the same work.

MM J Marchal & Cie, Saint Die (Vosges), France.—With former oil, were using 14.54 per cent more power for the same work.

Herren Methner & Frahne, Waldenburg, Ischl, Germany.—With former oil, were using 14.50 per cent more power for the same work.

Settsu Boseki Kabushiki Kwaisha, Osaka, Japan.—With former oil, were using 14.09 per cent more power for the same work.

Tsuchima Boseki Kabushiki Kaisha, Tsushima, Japan.—With former oil, were using 13.99 per cent more power for the same work.

Signori Ing Muggiani & Cia, Intra, Italy.—With former oil, were using 13.95 per cent more power for the same work.

Springfield Mill, Blackburn, England.—With former oil, were using 13.70 per cent more power for the same work.

Sunnyside Mill, Bolton, England.—With former oil, were using 13.60 per cent more power for the same work.

Société Cotonnière Mulhousienne, Mulhouse, Germany.—With former oil, were using 13.60 per cent more power for the same work.

Park Place Mill, Blackburn, England.—With former oil, were using 13.50 per cent more power for the same work.

Herren H P Geldermann & Söhne, Oldenzaal, Netherlands.—With former oil, were using 13.45 per cent more power for the same work.

Domenico Staurenghi, Carate Brianza, Italy.—With former oil, were using 13.27 per cent more power for the same work.

Spett Cotonoficio Veneziano, Venice, Italy.—With former oil, were using 13.22 per cent more power for the same work.

Grosvenor-Dale Co, Grosvenor-Dale, Connecticut, U S A.—With former oil, were using 13.15 per cent more power for the same work. Vacuum oils, all through, reduce the power, in one mill, from 1130 hp to 1000; in another, from 275 to 220; in another, from 455 to 400.

MM Dollfus & Cie, Belfort, France.—With former oil, were using 13.14 per cent more power for the same work.

Herren J J Rieter & Co, Nieder-Toss bei Winterthur, Switzerland.—With former oil, were using 13.12 per cent more power for the same work.

Filature & Filteries Réunies, Alos, Belgium.—With former oil, were using 12.74 per cent more power for the same work.

Spett Cotonificio Italiano, Forno (Massa), Italy.—With former oil, were using 12.58 per cent more power for the same work.

Spinnerei Forchheim G m b H (Carl Faber, Director), Forchheim, Bavaria.—Average of three tests—With former oil, were using 12.54 per cent more power for the same work.

A leading mill in Bury, England.—With former oil, were using 12.50 per cent more power for the same work.

Signor F E Mylius, Cogozzo (Brescia), Italy.—With former oil, were using 12.45 per cent more power for the same work.

Schonauer & Sollenauer Baumwollgarn-Manufactur, Sollenau, Austria.—With former oil, were using 12.42 per cent more power for the same work.

MM Doyen & Cie, Mondrepues (Aisne), France.—With former oil, were using 12.18 per cent more power for the same work.

Kanegafuchi Boseki Kabushiki Kaisha, Tokio, Japan.—With former oil, were using 12.07 per cent more power for the same work.

Herr Ernst Mauthner, Bubenč bei Prag, Austria.—With former oil, were using 12.07 per cent more power for the same work.

MM Etienne Motte & Cie, Roubaix (Nord), France.—With former oil, were using 12.07 per cent more power for the same work.

A Cotton Mill near Accrington, England.—With former oil, were using 12 per cent more power for the same work.

Anglo-Italian Hemp Spinning Co, Limited, Ferrara, Italy.—With former oil, were using 11.84 per cent more power for the same work.

Landeshuter Flachsgarn-Maschinenspinnerei, Landeshut, Schl.—With former oil, were using 11.69 per cent more power for the same work.

Herr Ferdinand Bohn, Flachsgarnspinnerei, Niederlangenau bei Hohenelbe, Netherlands.—With former oil, were using 11.63 per cent more power for the same work.

M Iwan Simonis, Verviers, Belgium.—With former oil, were using 11.61 per cent more power for the same work.

Herren Brevillier & C, Schwadorf, Austria.—With former oil, were using 11.37 per cent more power for the same work.

Jute-Mill of Signor P Frattini di L, Milan, Italy.—With former oil, were using 11.35 per cent more power for the same work.

Hronowere Baumwollspinnerei, Hronow, Bohemia.—With former oil, were using 11.23 per cent more power for the same work.

MM Boutemy Frères, Lannoy (Nord), France.—With former oil, were using 11.22 per cent more power for the same work.

Messrs William Barbour & Sons Limited, Lisburn, Ireland.—With former oil, were using 11.11 per cent more power for the same work.

A Cotton-Mill at Stalybridge, England.—With former oil, were using 10.81 per cent more power for the same work.

Herr Adolph Schwab, Hammerstein (Reichenberg), Austria.—With former oil, were using 10.50 per cent more power for the same work.

Gebrüder Walzel, Parschitz, Bohemia.—With former oil, were using 10.44 per cent more power for the same work.

Signor Gaetano Rossi di A, Vicenza, Italy.—With former oil, were using 10.34 per cent more power for the same work.

Société Anonyme pour la Filature des Laines Peignées, Tournai, Belgium.—With former oil, were using 10.15 per cent more power for the same work.

Signor P Mazzoni di G B, Pralafera (Torino), Italy.—With former oil, were using 10.13 per cent more power for the same work.

Signor Luigi Candiani, Garottola (Busto Arsizio), Italy.—With former oil, were using 9.85 per cent more power for the same work.

Herren Gebr Laurenz, Ochtrup, Westphalia, Germany.—With former oil, were using 9.20 per cent more power for the same work.

Herren von Velsen & Co, Bocholt, Germany.—With former oil, were using 8.83 per cent more power for the same work.

Société Cotonnière de St Etienne du Rouvray (Seine-Inférieure), France.—With former oil, were using 8.31 per cent more power for the same work.

Westdeutsch Jutespinnerei & Weberei, Beuel, Germany.—With former oil, were using 8.06 per cent more power for the same work.

Herr Johann Faltis Erban, Trautenau, Bohemia.—With former oil, were using 8 per cent more power for the same work.

Herren Josef Sobotka & Co, Lodenitz bei Prag, Austria.—With former oil, were using 7.60 per cent more power for the same work.

Messrs Wm Halley & Sons, Dundee, Scotland.—With former oil, were using 6.47 per cent more power for the same work.

Signori G Defferrari & Cia, Busalla, Italy.—With former oil, were using 4.34 per cent more power for the same work.

EXPLANATION OF DIAGRAM ON NEXT PAGE

The diagram on the next page shows the percentage results of the 111 tests related in nearly identical words on the preceding 18 pages.

The bottom line represents, for each one of the tests, the power required to do the work of that test with the Vacuum oil or oils.

The lines above represent, successively, one per cent more power than that required with the Vacuum oils in all the tests.

The dots represent the competing oils of all the tests. They show the results in this way: Each dot is on the percentage line that shows the result of its test.

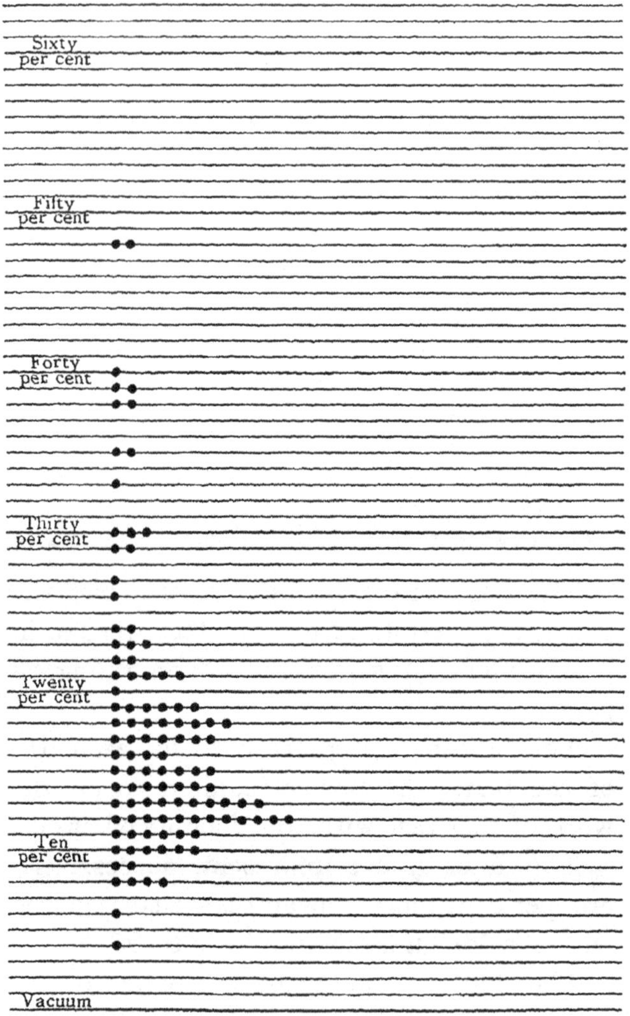

Diagram showing result of 111 tests
See page 67 for explanation

Thus: The dot on the 48-per-cent line represents the competing oil of the first of the 111 tests, at Termonde, Belgium—page 50. The loss of power was 48 per cent with the former oil of the mill.

And thus: The dot on the 4-per-cent line (fourth from bottom) stands for the last of the 111 tests, at Busalla, Italy—page 67.

The average gain in power is seen, on the diagram, to be from ten to twenty per cent.

These tests were mostly in spinning mills. A majority of them, perhaps, with Velocité spindle oil; a good many with Arctic spindle oil; a good many included Vacuum oils all through the mills.

The only way, to make such a diagram, is to throw together all the tests we have had, the data of which we have kept.

We may perhaps, in a future edition, revise the map, to include every possible test to date. In this, we have omitted all but one in a mill, and some that clearly ought to have been included.

It is almost true that no test has ever shown a less saving of power with Vacuum oils than with the competing oils; but that will occur, if we ever make a big mistake in the choice of an oil.

Bartlett & Company
The Orr Press
New York

Made in the USA
Monee, IL
04 November 2026

67388654R10046